THE ROLE OF NAVAL FORCES IN THE GLOBAL WAR ON TERROR

ABBREVIATED VERSION

Committee on the Role of Naval Forces in the Global War on Terror
Naval Studies Board
Division on Engineering and Physical Sciences

NATIONAL RESEARCH COUNCIL
OF THE NATIONAL ACADEMIES

THE NATIONAL ACADEMIES PRESS
Washington, D.C.
www.nap.edu

THE NATIONAL ACADEMIES PRESS 500 Fifth Street, N.W. Washington, DC 20001

NOTICE: The project that is the subject of this report was approved by the Governing Board of the National Research Council, whose members are drawn from the councils of the National Academy of Sciences, the National Academy of Engineering, and the Institute of Medicine. The members of the committee responsible for the report were chosen for their special competences and with regard for appropriate balance.

This study was supported by Contract No. N00014-05-G-0288, DO #3 between the National Academy of Sciences and the Department of the Navy. Any opinions, findings, conclusions, or recommendations expressed in this publication are those of the author(s) and do not necessarily reflect the views of the organizations or agencies that provided support for the project.

International Standard Book Number-13: 978-0-309-10179-0
International Standard Book Number-10: 0-309-10179-4

Additional copies of this report are available from:

Naval Studies Board, National Research Council, The Keck Center of the National Academies, 500 Fifth Street, N.W., Room WS904, Washington, DC 20001; and

The National Academies Press, 500 Fifth Street, N.W., Lockbox 285, Washington, DC 20055; (800) 624-6242 or (202) 334-3313 (in the Washington metropolitan area); Internet, http://www.nap.edu.

Printed in the United States of America

THE NATIONAL ACADEMIES

Advisers to the Nation on Science, Engineering, and Medicine

The **National Academy of Sciences** is a private, nonprofit, self-perpetuating society of distinguished scholars engaged in scientific and engineering research, dedicated to the furtherance of science and technology and to their use for the general welfare. Upon the authority of the charter granted to it by the Congress in 1863, the Academy has a mandate that requires it to advise the federal government on scientific and technical matters. Dr. Ralph J. Cicerone is president of the National Academy of Sciences.

The **National Academy of Engineering** was established in 1964, under the charter of the National Academy of Sciences, as a parallel organization of outstanding engineers. It is autonomous in its administration and in the selection of its members, sharing with the National Academy of Sciences the responsibility for advising the federal government. The National Academy of Engineering also sponsors engineering programs aimed at meeting national needs, encourages education and research, and recognizes the superior achievements of engineers. Dr. Wm. A. Wulf is president of the National Academy of Engineering.

The **Institute of Medicine** was established in 1970 by the National Academy of Sciences to secure the services of eminent members of appropriate professions in the examination of policy matters pertaining to the health of the public. The Institute acts under the responsibility given to the National Academy of Sciences by its congressional charter to be an adviser to the federal government and, upon its own initiative, to identify issues of medical care, research, and education. Dr. Harvey V. Fineberg is president of the Institute of Medicine.

The **National Research Council** was organized by the National Academy of Sciences in 1916 to associate the broad community of science and technology with the Academy's purposes of furthering knowledge and advising the federal government. Functioning in accordance with general policies determined by the Academy, the Council has become the principal operating agency of both the National Academy of Sciences and the National Academy of Engineering in providing services to the government, the public, and the scientific and engineering communities. The Council is administered jointly by both Academies and the Institute of Medicine. Dr. Ralph J. Cicerone and Dr. Wm. A. Wulf are chair and vice chair, respectively, of the National Research Council.

www.national-academies.org

COMMITTEE ON THE ROLE OF NAVAL FORCES IN THE GLOBAL WAR ON TERROR

MIRIAM E. JOHN, Sandia National Laboratories, *Co-chair*
RICHARD L. WADE, Exponent, *Co-chair*
H. NORMAN ABRAMSON, San Antonio, Texas
NOEL K. CUNNINGHAM, Glendora, California
KEVIN P. GREEN, IBM Global Business Services
RODNEY GREGORY, IBM Global Business Services
LEE HAMMARSTROM, Applied Research Laboratory, Pennsylvania State University
PAUL W. HOFF, Bedford, New Hampshire
JAMES D. HULL, Annapolis, Maryland
HARRY W. JENKINS, JR., Gainesville, Virginia
RONALD R. LUMAN, Applied Physics Laboratory, Johns Hopkins University
ANN K. MILLER, University of Missouri, Rolla
JOHN H. MOXLEY III, Solvang, California
GENE H. PORTER, Nashua, New Hampshire
JOHN S. QUILTY, Oakton, Virginia
BRAD ROBERTS, Institute for Defense Analyses
ANNETTE L. SOBEL, Sandia National Laboratories
H. EUGENE STANLEY, Boston University
MARLIN U. THOMAS, Air Force Institute of Technology
DAVID A. WHELAN, The Boeing Company
ELIHU ZIMET, Gaithersburg, Maryland

Staff

CHARLES F. DRAPER, Director
ARUL MOZHI, Study Director
EUGENE J. CHOI, Program Officer (through May 18, 2007)
IAN M. CAMERON, Associate Program Officer (through May 21, 2007)
SUSAN G. CAMPBELL, Administrative Coordinator
MARY G. GORDON, Information Officer
AYANNA N. VEST, Senior Program Assistant (through June 9, 2006)
SIDNEY G. REED, JR., Consultant
RAYMOND S. WIDMAYER, Consultant

Staff

CHARLES F. DRAPER, Director
ARUL MOZHI, Senior Program Officer
EUGENE J. CHOI, Program Officer (through May 18, 2007)
IAN M. CAMERON, Associate Program Officer (through May 21, 2007)
SUSAN G. CAMPBELL, Administrative Coordinator
MARY G. GORDON, Information Officer
AYANNA N. VEST, Senior Program Assistant (through June 9, 2006)

Preface

In December 2004 the Naval Studies Board was briefed on the Navy's Maritime Intercept Operations in support of the Proliferation Security Initiative, which seeks to stop the flow of weapons of mass destruction as cargo in the open sea. Given the historic roles of the Navy and the Marine Corps in conducting intelligence, surveillance, and reconnaissance (ISR) and offensive forward operations to deter and prevent terrorist actions against the homeland, this briefing was the genesis of the present study on the naval forces' role in the Global War on Terror.[1]

The United States is a maritime nation whose survival and economic vitality depend on the free flow of commerce, whether in energy, food, or consumer products. The maritime infrastructure abroad as well as at home is thus critical to U.S. strategic interests, particularly since more than 90 percent of U.S. trade moves by water. The Department of the Navy has defined for itself several missions for the Global War on Terror (GWOT): to establish foundations for cooperative interactions with other agencies and nations; to provide the ISR needed for integrated maritime surveillance; to protect U.S. forces and infrastructure; and to conduct maritime interdiction in areas beyond the U.S. Exclusive Economic Zone.[2]

The Department of the Navy has participated in GWOT military operations most heavily through the Marine Corps and the Navy's Special Operations Forces, providing close air support; conducting strike operations and facilitating

[1]The study's use of the terminology "Global War on Terror" is discussed in the Prologue.

[2]Chief of Naval Operations (ADM Michael Mullen, USN). 2005. *CNO Guidance for 2006: Meeting the Challenge of a New Era,* Department of the Navy, Washington, D.C., October 30; Chief of Naval Operations (ADM Michael G. Mullen, USN). 2006. *Navy Strategic Plan in Support of Program Objective Memorandum 08,* Department of the Navy, Washington, D.C., May.

maneuvers in theater; providing other support such as explosive ordnance disposal and work on improvised explosive devices; establishing a Marine Corps special brigade and the Chemical and Biological Incident Response Force; and operating the National Maritime Intelligence Center.

The importance of maritime activities and vulnerabilities to the GWOT, particularly with respect to homeland defense, has been increasingly recognized in recent directives. The Department of Defense (DOD) has given the U.S. Northern Command (NORTHCOM) responsibility for military aspects of homeland defense. The maritime component commander for NORTHCOM is the Commander of the Fleet Forces Command, and the Navy's Third Fleet provides the Joint Force Maritime Component. The NORTHCOM Katrina Task Force was an example of the military's involvement in humanitarian operations, and it demonstrated how the military can contribute to recovery from a large-scale disaster. Although that task force was not set up in response to the GWOT, it embodies an approach that might help deter terrorist activity. Naval forces deployed to support tsunami disaster relief have demonstrated how humanitarian assistance and disaster relief operations overseas can help to dampen anti-American sentiment abroad.

Maritime security aspects of the GWOT have recently received top-level emphasis as national, DOD, and Service strategies have been drawn up.[3] Though the Navy has been slower than the other Services to step up to the GWOT challenges, it is starting to realign priorities to catch up. Its greatest challenge is building global maritime domain awareness (MDA) in order to prosecute the GWOT as far forward as possible. At the same time, it must be kept in mind that the attacks of 9/11, and earlier on the USS *Cole*, were executed from land. It must also be recognized that the maritime aspects of the GWOT involve operations with short notice, and that there is heavy dependence on command, control, communications, computers, intelligence, surveillance, and reconnaissance (C4ISR) to provide MDA that is adequate to support timely decisions on naval force allocations and dispositions for conducting a range of maritime operations.

In response to a request from the former Chief of Naval Operations,[4] the Naval Studies Board through the National Research Council (NRC) established the Committee on the Role of Naval Forces in the Global War on Terror to conduct an assessment of the adequacy of and prospects for improving the role of

---

[3]White House (George W. Bush), 2005, *The National Strategy for Maritime Security,* Washington, D.C., September (DOD and the Department of Homeland Security (DHS) are developing eight supporting implementation plans); Office of the Chairman, Joint Chiefs of Staff (Gen Peter Pace, USMC), 2006, *National Military Strategic Plan for the War on Terrorism,* Joint Chiefs of Staff, Washington, D.C., February 1.

[4]ADM Vern Clark, USN, CNO, letter dated April 18, 2005, to Dr. Bruce Alberts, President, National Academy of Sciences.

Naval Forces in the Global War on Terror. The committee's approach in responding to the study's terms of reference is outlined below.[5]

THE COMMITTEE'S APPROACH IN RESPONDING TO THE TERMS OF REFERENCE

The Committee on the Role of Naval Forces in the Global War on Terror (see Appendix B for biographies of committee members) convened in July 2005 and held meetings over a period of 6 months to gather input from the relevant communities and then to discuss the committee's findings (summarized agendas of the meetings are provided in Appendix C).[6] The months between the committee's last meeting and the publication of the report were spent drafting the manuscript, gathering additional information, reviewing and responding to external review comments, editing the report, and conducting the security/public release review required to produce this version of the report that does not disclose information as described in 5 U.S.C. 552(b). It was mutually determined by the Department of the Navy and the National Research Council that the full report contained information as described in 5 U.S.C. 552(b) and therefore could not be released to the public in its entirety.

The initial approach of the committee was to identify specific operational, policy, and technical areas necessary to fulfill the tasks in the terms of reference. However, briefings soon revealed that Navy thinking and activities related to the GWOT were continuing to evolve. For example, while most of the topics in the terms of reference focus on operational and technical capabilities for maritime security operations, the *CNO Guidance for 2006* reflected a much broader perspective on the GWOT.[7] The committee therefore believed that to carry out its charge to assess the adequacy of and prospects for improvement of the role of naval forces in the GWOT, it had to adopt a broad interpretation of the terms of reference.

Lacking a Naval Services-generated framework comprehensive enough to delineate the spectrum of threats, environments, and missions pertinent to naval forces, the committee developed a Defense-in-Depth framework (see Figure ES.1 in the Executive Summary) to address the issues of operational and technical capabilities called for in the terms of reference and by the CNO guidance. The committee spent considerable time debating the approach and then developing the framework as the organizing construct for assessing needed capabilities, status, and gaps. Its members came to believe that the Navy must do the same in a "top-down" fashion that maps the full problem and mission space across the

[5]The terms of reference for this study are listed in Appendix A.

[6]During the entire course of its study, the committee held meetings in which it received (and discussed) materials that are exempt from release under 5 U.S.C. 552(b).

[7]Chief of Naval Operations (ADM Michael Mullen, USN). 2005. *CNO Guidance for 2006: Meeting the Challenge of a New Era,* Department of the Navy, Washington, D.C., October 30.

continuum of GWOT operations. Such a framework would address the problem end-to-end by integrating operations in disparate theaters of warfare into a continuous maritime whole.

Furthermore, the Defense-in-Depth framework was developed by the committee to better convey the messages that attach to the naval forces' GWOT mission. Because the committee's examination of the principal elements of the Defense-in-Depth framework resulted in findings and recommendations that could be grouped into seven areas for priority action, the Executive Summary summarizes the recommendations under these seven priority areas. Finally, the committee emphasizes the Navy's roles and missions in the GWOT because the Navy has trailed the other Naval Services in this area, but due attention to activities of the Marine Corps and the Coast Guard has been given where appropriate.

ACKNOWLEDGMENTS

The Committee on the Role of Naval Forces in the Global War on Terror thanks the many briefers who presented information essential to the writing of this report. Special acknowledgment goes to RADM Samuel J. Locklear III, USN, and RDML Dan W. Davenport, USN, and their staff at the Assessment Division, Office of the Deputy Chief of Naval Operations for Resources, Requirements, and Assessments (N81), who helped the committee coordinate numerous briefings and facilitated information gathering throughout this study. The committee also thanks VADM Kevin Cosgriff, USN, and his staff for hosting the committee and arranging for briefings during the meeting on November 1-2, 2005, at the U.S. Fleet Forces Command, and CAPT Karl Heinz, USN, and his staff for hosting the committee and arranging for briefings during the meeting on September 26-27, 2005, at the Naval Special Warfare Command.

The committee acknowledges and thanks ADM Dennis C. Blair, USN (retired), for his valuable insights while he was a member of the committee during the drafting of the report. Finally, the committee thanks the dedicated staff of the Naval Studies Board: Dixie Gordon, who facilitated the handling of the information associated with this study; Susan Campbell, who capably saw to the logistics for our many meetings over a short period of time and helped with report production; Ian Cameron, who added to the exceptionally smooth conduct of the meetings and the production of the report; Eugene Choi, who supported the response to report review; Sidney Reed, who supported report development and review through document research and cogent observations; Raymond Widmayer, who compiled the meeting agendas and arranged for site visits; and, in particular, our study director, Arul Mozhi, who kept the study on track and provided invaluable support to us and to the committee members.

Miriam E. John and Richard L. Wade, Co-chairs
Committee on the Role of Naval Forces in the Global War on Terror

Acknowledgment of Reviewers

National Research Council (NRC) reports are reviewed in draft form by individuals chosen for their diverse perspectives and technical expertise, in accordance with procedures approved by the NRC's Report Review Committee. The purpose of this independent review is to provide candid and critical comments that will assist the institution in making its published reports as sound as possible and to ensure that the reports meet institutional standards for objectivity, evidence, and responsiveness to the study charge. The review comments and draft manuscripts remain confidential to protect the integrity of the deliberative process. Although the reviewers provide many constructive comments and suggestions, they are not asked to endorse the conclusions or recommendations nor do they see the final draft of reports before release. We wish to thank the following individuals for their review of the draft report:

Lewis M. Branscomb, Harvard University,
RADM Erroll Brown, USCG (retired), IBM Global Business Services,
Ruth A. David, ANSER,
Seymour J. Deitchman, Chevy Chase, Maryland,
Gen Alfred M. Gray, USMC (retired), The Potomac Institute for Policy Studies,
Catherine M. Kelleher, University of Maryland,
Jerry A. Krill, Applied Physics Laboratory, Johns Hopkins University,
ADM Donald L. Pilling, USN (retired), LMI,
Edward Wenk, Jr., University of Washington, and
George M. Whitesides, Harvard University.

The review of the draft report was overseen by Alexander H. Flax, Potomac, Maryland. Appointed by the NRC, he was responsible for making certain that an independent examination of the draft report was carried out in accordance with institutional procedures and that all review comments were carefully considered. Responsibility for the final content of NRC reports rests entirely with the authoring committee and the institution.

Contents

Prologue

In response to the attacks of September 11, 2001, the United States has sought to expand and transform tools of policy and the roles of government agencies to confront and prevail against terrorist threats. Since 9/11, the nation has engaged in what national leadership has termed the "Global War on Terror" (GWOT). Its current objectives are outlined in the latest national security strategy of the United States:[1]

- Prevent attacks by terrorist networks before they occur.
- Deny WMD [weapons of mass destruction] to rogue states and to terrorist allies.
- Deny terrorist groups the support and sanctuary of rogue states.
- Deny terrorists control of any nation they would use as a base and launching pad for terror.

The *National Military Strategic Plan for the War on Terrorism* calls out the principal military objectives:[2]

- Protect and defend the homeland.
- Attack terrorists and their capacity to operate effectively at home and abroad.
- Support mainstream Muslim efforts to reject violent extremism.

---

[1]The White House (George W. Bush). 2006. *The National Security Strategy of the United States of America,* Washington, D.C., March, p. 12.

[2]Office of the Chairman, Joints Chiefs of Staff (Gen. Peter Pace, USMC). 2006. *National Military Strategic Plan for the War on Terrorism*, Joint Chiefs of Staff, Washington, D.C., February 1, p. 3.

That plan also identifies ways to expand foreign partnerships and partnership capacity; strengthen capacity to prevent terrorist acquisition and use of WMD; and institutionalize domestically and internationally the strategy against violent extremists. The GWOT involves multiple operations for the U.S. military, from combating terrorist threats to counterproliferation to providing humanitarian assistance.[3]

In meeting these new and growing demands that it contribute to the nation's GWOT efforts, the U.S. military faces a strategic landscape that has changed dramatically over the last two decades and that is likely to evolve for the foreseeable future.[4] A complicating factor in that evolution is the fact that the GWOT is not a "war" per se. It is a complex conflict involving many interdependent aspects of political, diplomatic, economic, and military policy. The Committee on the Role of Naval Forces in the Global War on Terror acknowledges that there has been much public discussion of whether the phrases "war on terror" and "Global War on Terror" are appropriate descriptions,[5] given the lack of an organized, uniformed, specified hostile force like that experienced in conventional campaigns. However, history has shown that past conflicts have taken on the vernacular label "war" in its most euphemistic sense. For example, the Cold War was not technically a war but an ideological, geopolitical, and economic struggle based on containment and deterrence policies that lasted for nearly half a century. While it came to be focused on the Soviet Union in its later phases, in its early years the Cold War shared a number of commonalities with the GWOT and where we stand in relation to it today:

- The problem we faced then, as now, was extraordinarily difficult.
- We had a limited understanding of our adversaries.
- Technology was evolving rapidly and was available to both sides.
- The possibility of catastrophic attacks was non-negligible.
- The confrontation was global.

Of course, the specifics of the GWOT differ and seem more complex. In the Cold War, two nominally symmetrical militaries faced each other in a prolonged stand-off; in the GWOT, myriad asymmetric adversaries are involved in a war that is to varying degrees warm and hot. In the Cold War, deterrence was central; in the GWOT, deterrence seems far less promising (but has also been far

---

[3]See, for example, the White House (George W. Bush), 2003, *Progress Report on the Global War on Terrorism*, Washington, D.C., September; Secretary of Defense (Donald H. Rumsfeld), 2006, *Quadrennial Defense Review Report*, Department of Defense, Washington, D.C., February 6. See Box 1.1 in Chapter 1 for a complete list of documents relating to national security strategy.

[4]Challenges facing the U.S. military are homeland defense, the GWOT and irregular warfare, and conventional campaigns. See Secretary of Defense (Donald H. Rumsfeld), 2006, *Quadrennial Defense Review Report*, Department of Defense, Washington, D.C., February 6.

[5]For example, see William Safire, "Islamofascism," *New York Times*, October 1, 2006, p. 20.

less studied). Adversary attacks in the GWOT involve both civilian and military targets, and terrorists are not deterred by threats to their survival. Their depth of military capability is not great, but they have demonstrated opportunism and persistence, and they seem to be able to adapt so as to cause effects disproportionate to their military strength. Multiple religious and ideological agendas motivate their efforts. The agendas are mixed with fighting not only against the United States, Western (e.g., Britain, Italy, Spain), Middle Eastern (e.g., Iraq, Saudi Arabia), and Asian (e.g., India, Indonesia, the Philippines, Thailand) interests, but also within their own factions. The terrorists' efforts are apparent at levels that range from isolated cells, through blurred organizations such as al-Qaeda, to nation-states that supply help ranging from passive sanctuary to active support. Terrorist actions and areas of influence—from Afghanistan and Iraq to critical maritime straits such as Malacca and Hormuz, and to areas holding most of the known petroleum reserves—are also broad. Organizations like al-Qaeda have considerable resources, own ships, use technologies and principles of network-centric operations, and have made serious efforts to obtain WMD. This complex GWOT environment portends a long-term conflict that has extensive maritime dimensions.

The committee saw its charter as being neither to endorse nor to replace the term "GWOT," but rather to describe the scope of the problem to which naval forces should be directed. In the committee's view, the GWOT refers to national-security-related conflict, offensive operations, and defense tied directly to blunting terrorists threats, whether they are state-sponsored or not. But the committee believes as well that the long-war scope of the GWOT must also be addressed in terms of operations beyond combating terrorism, such as counter-proliferation, strengthening alliances, and providing humanitarian assistance and disaster relief—activities that build a worldwide base for containing and prevailing against terrorism. It is this broader view of the GWOT that the committee has taken in defining roles, missions, and needed capabilities for naval forces.[6]

[6]In this report, the committee focused on the role of the Navy in the GWOT. In several areas the inter-related roles of the Marine Corps and the Coast Guard are also addressed. Thus, the term "naval forces" as used here refers to the Navy, the Marine Corps, and the Coast Guard.

Executive Summary

At the request of the former Chief of Naval Operations (CNO), the Naval Studies Board, under the auspices of the National Research Council, established a committee to assess the capabilities and gaps of naval forces in prosecuting the Global War on Terror (GWOT).[1] The Committee on the Role of Naval Forces in the Global War on Terror was established late in the tenure of ADM Vern Clark, USN, the previous CNO, and conducted its work in the first several months of the tour of the current CNO, ADM Michael G. Mullen, USN. During that period, the Navy began to shift priorities toward addressing naval roles in the GWOT. As a result, the findings and recommendations of the committee often support activities recently set in motion. Many areas remain to be addressed, however, because of the breadth and complexity of the GWOT, the interrelationships of political, diplomatic, economic, and military policies and operations, and the fact that the intellectual work to date has not matched the breadth and complexity of their interrelationships. These shortfalls cannot be dismissed. The GWOT is expected to be a long war, and there will be an enduring requirement for naval operations.

[1]In the committee's view, the GWOT refers to national-security-related conflict, offensive operations, and defense tied directly to blunting terrorist threats, whether they are state-sponsored or not. But the committee believes as well that the long-war scope of the GWOT must also be addressed in terms of operations beyond combating terrorism, such as counterproliferation, strengthening alliances, and providing humanitarian assistance and disaster relief—activities that build a worldwide base for containing and prevailing against terrorism. It is this broader view of the GWOT that the committee has taken in defining roles, missions, and needed capabilities for naval forces. In this report, the committee focuses on the role of the Navy in the GWOT. In several areas the interrelated roles of the Marine Corps and the Coast Guard are also addressed. Thus, the term "naval forces" as used in this report refers to the Navy, the Marine Corps, and the Coast Guard.

The charge to the committee, as stipulated in the terms of reference (see Appendix A), was therefore broad and complex. Lacking a comprehensive construct from any of the Naval Services that would allow a systematic assessment of capabilities, status, and gaps, the committee found it necessary to develop a Defense-in-Depth framework that became the organizing construct for this report. The committee's examination of the principal elements of the framework resulted in findings and recommendations that could be grouped into seven priority areas for action. This summary assesses the transformation of naval forces for addressing the GWOT as of the writing of this report, briefly describes the Defense-in-Depth framework, and lists the findings and associated major recommendations for each action area.

NAVAL TRANSFORMATION FOR THE GWOT

For the U.S. military the strategic landscape has changed dramatically over the last two decades. As national strategy documents have argued, a central feature of this new landscape is uncertainty—about the state or nonstate adversaries that might threaten U.S. security, about their capabilities, and about their intentions. This uncertainty has profound implications for naval planners. They cannot transform naval forces for the GWOT without worrying about how such transformations will work for or against the other transformations needed for non-GWOT challenges (for example, major combat operations [MCOs]). Nor can they tailor the GWOT strategy for a clearly identifiable source or type of terrorist action. Naval planners must link capability development plans to the objectives set by national leadership. Despite uncertainties about the future of this long war, force planners must promote the kinds of adaptations political leaders seek. They must also take a long-term view of the problem and of the solution. This can be, but is not likely to be, a traditional force-on-force application of naval seapower.

On the good news side of the ledger, Coast Guard leadership has embraced transformation for the GWOT, and the Marine Corps has moved aggressively to bolster capabilities for GWOT-relevant missions. The CNO's 2006 guidance[2] puts the GWOT at the top of the warfighting priority list. The CNO has followed with taskings and decisions consistent with that priority.[3] He has reinforced these taskings and decisions with a strategic plan in support of Program Objective Memorandum 08 that lends detail and commitment to such initiatives.[4] Fleet commanders are starting to define needed capabilities through their GWOT-

---

[2]Chief of Naval Operations (ADM Michael Mullen, USN). 2005. *CNO Guidance for 2006: Meeting the Challenge of a New Era,* Department of the Navy, Washington, D.C., October 30.

[3]For example, standup of the Naval Expeditionary Command; standup of the riverine unit; expansion of the foreign affairs officer program; and expansion of explosive ordnance disposal (EOD) capabilities.

[4]Chief of Naval Operations (ADM Michael G. Mullen, USN). 2006. *Navy Strategic Plan in Support of Program Objective Memorandum 08,* Department of the Navy, Washington, D.C., May.

related interactions with combatant commands and, in the case of the Pacific Fleet, with special attention to operations inherent to forward presence.

However, of the three principal Naval Services—the Navy, the Marine Corps, and the Coast Guard—the Navy's record of innovation and adaptation to address the GWOT challenge is the least advanced in terms of the expectations of national leadership.[5] In the committee's view, its record also falls short of the likely future requirements for success. Hence, this report focuses primarily on the role of the Navy in the GWOT.

The most urgent challenge is the question of future requirements. In the committee's view, the Navy will not have done this job rigorously unless it integrates and balances the demands of the war on terror and irregular warfare, homeland defense, and conventional campaigns. It must also understand the separate requirements of improved performance in the steady state, as the new *Quadrennial Defense Review Report* defines it,[6] and the surge phase. Moreover, given the requirements of maritime security in the new strategic environment, the Navy must become an effective player in, and a leader of, the interagency process bearing on these matters.

DEFENSE-IN-DEPTH FRAMEWORK

A basic tenet of this report—and of national leadership—is that the GWOT will be a long war. Naval forces, particularly the Navy, must therefore institutionalize mechanisms for treating it as an enduring mission. One such mechanism is to create the intellectual base for assessing requirements for the GWOT on its own terms to allow sound prioritization of options, both within the GWOT and across the full mission space of naval responsibilities. The committee believes that an intermediate level of detail, one that bridges the gap between strategic commitment and specific measures that intuitively provide capabilities and capacities for the GWOT, is needed. To highlight this point and to provide an organizing construct for its own assessment, the committee drew up the Defense-in-Depth framework illustrated in Figure ES.1.

Three roles for naval forces are critical to forming a defense in depth against GWOT threats,[7] operating in and from the maritime domain, including along U.S. coasts. They are forward presence, maritime operations, and homeland defense. There are also three critical foundational capabilities necessary to support naval forces engaged in the GWOT: maritime domain awareness (MDA), which includes increased maritime intelligence; command and control; and naval

[5] A discussion of capability gaps in the GWOT is summarized in the section titled "Develop Naval Strategy."

[6] Secretary of Defense (Donald H. Rumsfeld). 2006. *Quadrennial Defense Review Report,* Department of Defense, Washington, D.C., February 2, p. 4.

[7] It is recognized that threats can originate anywhere geographically, and in cyberspace as well as physically.

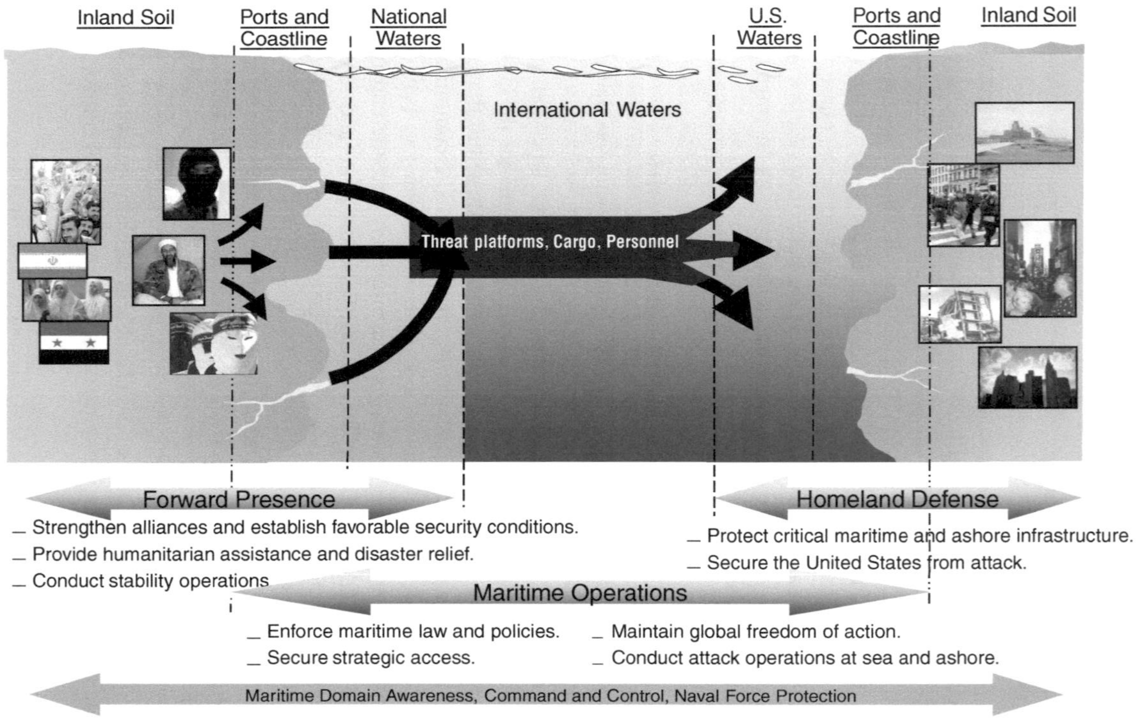

FIGURE ES.1 Defense-in-Depth framework for the Global War on Terror in the maritime domain.

force protection. Finally, three implementing areas crosscut the GWOT roles and capabilities: (1) technologies unique to the maritime domain; (2) specialized fleet capabilities tailored to the dispersed and elusive nature of the threat; and (3) new, or newly emphasized, personnel skills and training.

PRIORITY AREAS FOR ACTION

The committee developed its findings and recommendations based on its assessment of each of the elements that constitute the Defense-in-Depth framework, but recognized that the breadth of the topic and the many recommendations that resulted did not provide a useful starting point for Service leadership. It therefore identified seven areas for priority action to which the major and supporting recommendations could be mapped:

1. *Develop naval strategy,* both top-down as part of the national strategy and bottom-up based on naval capabilities, to address the naval role in the GWOT—and that strategy's derivatives in terms of the concept of operations (CONOPS), capabilities, and investment—in its own terms.
2. *Assign responsibilities* for work on the naval approach to the GWOT, including identifying valid requirements, inventorying current and programmed capabilities, and deciding on investment priorities and resource commitments.
3. *Strengthen maritime domain awareness* and the spectrum of options to deal with the fact that the maritime domain represents an all-too-plausible channel for delivery of terrorists and WMD, especially nuclear.
4. *Seize opportunities for forward presence* to leverage this traditional naval strength relative both to winning hearts and minds and to conducting operations with foreign partners.
5. *Prioritize assets for increased protection* to include not only homeland population centers but also homeland and overseas critical infrastructure[8] and the forces themselves.
6. *Operationalize the Navy/Coast Guard "national fleet" concept* as a centerpiece of the national maritime security strategy, recognizing that the complementary strengths of partner nations are especially valuable in the GWOT (consistent with the motivation of the emerging "1,000-ship Navy" concept),[9] and planning for GWOT-unique fleet capabilities in the future.
7. *Attract, develop, and retain the right people,* especially more diversified

[8]Recognizing that certain critical economic infrastructure represents maritime targets that have strategic consequence; these warrant increased monitoring as part of maritime security.

[9]The "1,000-ship Navy" concept foresees an international, interoperable coalition of naval-related activities, joined by nations allied to or friendly with the United States. The concept further proposes a vastly expanded sensor network to monitor security in the maritime domain and an increased number of responders helping to ensure this security. Ultimately, the objective is to allow the maritime domain to be safely used by all cooperating nations.

special operations personnel and skilled specialists who are attuned to local and cultural norms in forward deployed areas—people who can build the enduring personal and professional relationships that are at the heart of creating an effective GWOT force.

Develop Naval Strategy

Finding: The Navy's strategic plan of May 2006[10] makes good use of the new force planning construct in the Quadrennial Defense Review (QDR) to structure its discussion of focus areas and mission sets, but it needs to exploit that construct to assemble a vision of how to prevail (one that goes beyond what to do) in the GWOT. Such a vision can help motivate a broad set of activities and illuminate the gaps in current capabilities as this committee has come to understand them.

Specific concerns of the committee about the Navy's May 2006 strategic plan are as follows:

- The strategic plan is too "Navy" and not enough "naval." The Navy plan conveys only a modest appreciation of how to integrate a Navy, Marine Corps, and Coast Guard team to prosecute the GWOT.
- The concept of joint forces reflected in Navy strategy is too constrained. In the new strategic landscape, jointness expands to include coalition operations. It must encompass the full spectrum of partner capacities. Motivating partner capacities is an important function of strategy as reflected, for instance, in the emerging concept of a 1,000-ship Navy.
- The Navy strategy is generally written in the language of "defend" and "defeat." It needs to reflect the objectives of assurance, dissuasion—and especially deterrence. Reassuring U.S. partners is central to enlisting their participation in the capability-building effort; that requires more than simply strength and presence. Dissuading potential future challengers, which is central to prevailing in complex contingencies, requires more than simply a notion of military victory. As national and DOD strategy documents point out, deterrence is especially important. Partners are deeply concerned with deterrence—they want to know how improvements in operational art and incremental capability can make a meaningful difference in terrorist behavior. While the Navy's May 2006 strategic plan acknowledges this, the strategy gives no answers to how the programmatic guidance connects to this critical objective.
- There is too little focus in the strategy on the WMD threat posed by terrorists. The Navy's May 2006 strategic plan makes a significant commitment to countering the WMD threat, but the contribution of particular measures is not yet

---

[10]Chief of Naval Operations (ADM Michael G. Mullen, USN). 2006. *Navy Strategic Plan in Support of Program Objective Memorandum 08,* Department of the Navy, Washington, D.C., May.

clear because substantive analyses like those conducted for MCO capabilities and capacities have not yet been done.

- Combatant commander (COCOM) priorities are not reflected in the Navy Strategic Plan. Yet the COCOMs have some very clear ideas about shortfalls in current naval capabilities and capacities, and their preferences need to be explicitly reflected in the Navy's view of its emerging requirements.

Major Recommendation 1: The Chief of Naval Operations should develop and promulgate a strategy for the GWOT that begins with first principles and then explains how present and future naval forces can be organized, trained, equipped, and positioned to achieve national objectives. The required vision should:

- Begin with a view of the integrated nature of the "one game"; it is not a "home game" and an "away game"—but "one game."
- Articulate synergies among the three missions of the QDR.
- Encompass the combined roles of the nation's naval assets.
- Adopt a broader view of the Joint Force.
- Address the deterrence question more thoroughly in terms of the values and nature of the opposition.
- Address the WMD threat more effectively.
- Reflect combatant commander priorities.
- Build a firm foundation from the national strategy architecture.
- Develop GWOT planning scenarios that are analogous to MCOs that guide planning for conventional warfare capabilities and capacities, incorporating concepts of assurance, dissuasion, and deterrence as well as defense and defeat.

Assign Responsibilities

Finding: The Navy and its Marine Corps and Coast Guard partners have not defined the requirements of success in the GWOT in the GWOT's own terms. The committee believes that this will require a good deal of intellectual work, which is just beginning.[11] As stated above, the committee believes that an intermediate level of detail is needed to bridge the gap between strategic commitment and specific measures that intuitively provide capabilities and capacities for the GWOT. The Navy does not yet have such an operational construct, nor has it assigned responsibility for advocacy of the roles and requirements for naval forces in the GWOT needed to improve capabilities.

In association with this finding, the committee makes three interrelated recommendations.

---

[11] See, for example, Department of the Navy, 2005, *Navy's 3/1 Strategy: The Maritime Contribution to the Joint Force in a Changed Strategic Landscape,* Washington, D.C., April 12 (draft); Chief of Naval Operations (ADM Michael G. Mullen, USN), 2006, *Navy Strategic Plan in Support of Program Objective Memorandum 08,* Department of the Navy, Washington, D.C., May.

Major Recommendation 2: The Chief of Naval Operations should use the force planning construct in the 2006 Quadrennial Defense Review[12] to size and shape its forces for both the GWOT and MCO. This requires coming to terms with the different requirements of steady-state and surge, as defined by the QDR, and with the need to improve institutional capabilities to compete with an adaptive adversary.

Major Recommendation 3: The Chief of Naval Operations should develop or adopt an operational framework similar to the committee's Defense-in-Depth framework to structure force planning and analyses of the value-added of various candidate capabilities and capacities, and should assign advocacy responsibility to the Deputy Chief of Naval Operations for Integration of Capabilities and Resources (N8) for each segment of the framework.[13] The support of NWDC and MCCDC should be enlisted. (The foundational MDA capability is specifically called out in Major Recommendation 5.)

Major Recommendation 4: As part of the Navy accepting the leadership role required to support the maritime GWOT missions, the Deputy Chief of Naval Operations for Integration of Capabilities and Resources (N8) should address capability needs and technology shortfalls. To this end, the CNO should assign advocacy responsibility within N8 and support the development of assessment methods that will allow prioritization of GWOT investments, both among themselves and relative to other mission needs. Associated maritime GWOT planning scenarios should also be developed analogous to the MCOs that guide planning for conventional warfare capabilities and capacities. Again, the support of NWDC and MCCDC should be enlisted.

Finding: The Navy is not providing MDA leadership or institutional commitment commensurate with its responsibility—and its unique capabilities—to aggressively support the achievement of national MDA goals. The Navy has interpreted its MDA responsibilities narrowly in terms of responses to intelligence cues about vessels of interest. This has led to a focus on the processing of sensor data, primarily from the National Maritime Intelligence Center, and not on the need for new sensor capabilities as called for in the *National Plan to Achieve Maritime Domain Awareness*.[14] The current narrow approach neglects the larger

---

[12]Secretary of Defense (Donald H. Rumsfeld). 2006. *Quadrennial Defense Review Report,* Department of Defense, Washington, D.C., February 2, p. 38.

[13]The identification of specific officers and offices in the Navy with specific recommended actions is intended to reflect those most closely aligned in terms of the existing structures of organizational responsibilities.

[14]Department of Homeland Security. 2005. *National Plan to Achieve Maritime Domain Awareness for the National Strategy for Maritime Security,* Washington, D.C., October. Available at <http://www.dhs.gov/interweb/assetlibrary/HSPD_MDAPlan.pdf>. Accessed September 21, 2006.

challenge of leading the needed international effort to establish an active ocean and foreign port surveillance system that would help identify potential threats in time to permit an effective response. Such an MDA active surveillance capability is much more ambitious than that needed to protect naval forces against terrorists or indeed, to prosecute most major combat operations.

Major Recommendation 5: The Chief of Naval Operations should assign clear responsibility and accountability for maritime domain awareness within the Navy and direct the elevation of Navy representation and leadership within the interagency domain—as required in *The National Strategy for Maritime Security*.[15]

"Within the Navy" implies the clear designation of resource and planning responsibility within the Office of the Chief of Naval Operations (OPNAV). More broadly, charting an MDA architecture and roadmap—with specific attention to filling gaps—is essential.

The senior Navy officer responsible for interagency representation not only should take an active leadership role within the interagency arena but also should ensure that the Navy input adheres both to its own and to interagency capability and program processes and deliberations.[16,17]

Strengthen Maritime Domain Awareness

Finding: There are serious MDA capability gaps in terms of plausible, difficult maritime scenarios involving WMD threats to the United States and high-consequence threats to its economic infrastructure. These gaps call for considerable effort to explore and prioritize among solution options. Moreover, analytic processes are minimally automated and are challenged by large volumes of information for analysis and fusion.

Major Recommendation 6: The Deputy Chief of Naval Operations for Integration of Capabilities and Resources (N8) and the Assistant Secretary of the

[15]White House (George W. Bush). 2005. *The National Strategy for Maritime Security,* Washington, D.C., September.

[16]A Navy Maritime Domain Awareness–Implementation Team (MDA–IT) was being formed as of this writing, paralleling the interagency MDA–IT and offering the promise of more coordinated and coherent MDA efforts both within the Navy and as input to the interagency effort.

[17]*The Politics of the Oceans* (Edward Wenk, Jr., 1972, University of Washington Press, Seattle) gives the detailed perspective of a key White House participant on how the interagency, national, and international negotiations were conducted that resulted in government reorganization to form NOAA and EPA and in legislation on the status of the sea bottom. During the very active decade of the 1960s, many combinations of scientific, technological, economic, and political interests were orchestrated to generate institutions for investigation, use, and protection of the oceans. Perceptions of the major economic potential of the sea bottom were developed, leading to international political interactions. The book contains accounts of activities successful and unsuccessful, which could be useful to those carrying out this recommendation.

Navy for Research, Development, and Acquisition (ASN[RDA]) should co-lead a robust Navy-wide analytic and engineering effort to specify system and capability options to address high-stakes MDA threats. An engineering analysis team should be convened to develop a quantitative analytic methodology that maps MDA attributes to mission effectiveness for GWOT-related reference scenarios. The results should be used as a basis for Navy, Coast Guard, and related interagency decisions and actions. In addition, if successful in forming the MDA architecture and implementation decisions, this methodology should be extended and applied to a full defense-in-depth framework, either the one that is offered in this report or its replacement.

There is no single program or technology "silver bullet" solution. The recommended effort would address the MDA enterprise/system-of-systems options and the attendant cost, risk, and schedule considerations, along with the performance of different combinations of sensor/surveillance building blocks. These building blocks would include non-Navy capabilities and assets, both current and programmed or potential.

Finding: To yield an integrated national capability, further coherence is needed among the many community initiatives for MDA and MDA-supporting concepts.

Major Recommendation 7: The Chief of Naval Operations, the Commandant of the Coast Guard, and the Commander of the Northern Command (NORTHCOM) should co-sponsor a Navy, Coast Guard, and NORTHCOM effort to address coherence by developing a maritime domain awareness enterprise operational architecture (if this has not already been done by the time this report is issued).

Independent of the methodology employed, an operationally driven MDA architecture, national in scope, is needed. Such an architecture can in some sense be viewed as a more fleshed out version of the MDA CONOPS being developed by the interagency MDA–Implementation Team. The recommendation that this be accomplished by a chartered joint but not full interagency effort is based on the pragmatic recognition that the Navy, the Coast Guard, and NORTHCOM are the principal owners of the topic and the principal stakeholders in a coherent outcome. Their joint product would both ensure coherence among their core efforts and help drive the interagency process.

Seize Opportunities for Forward Presence

Finding: Humanitarian support can have a positive and highly leveraged impact on how the United States and its citizens are viewed by foreign populations, which would help improve the maritime security environment and further GWOT objectives.

Finding: Naval forces—especially the Navy—are not yet approaching forward presence as a strategic aspect of the GWOT. The Navy needs to implement a deliberate and comprehensive approach to both planning and engagement to maximize the impact of forward presence in the GWOT.

Finding: More effort and resources are needed to build Navy capabilities to contribute fully to enhanced engagement and theater security cooperation (TSC) efforts in support of COCOM requirements with forward-deployed naval forces. Recent priorities have focused on military-to-military relationships to enable TSC operations; they have not leveraged other public agencies or the private sector. (Recent initiatives by the Pacific Fleet are an exception.) Using the Navy, the Coast Guard, or other agencies with the right tools to gain access and then leverage future naval focus is critical to long-term success.

Major Recommendation 8: To shape the maritime security environment, in concert with COCOM TSC plans related to forward-deployed naval forces, the Deputy Chief of Naval Operations for Information, Plans, and Strategy (N3/N5) and the Deputy Commandant of the Marine Corps for Plans, Policies, and Operations (DCMC[PPO]) should coordinate with naval component commanders to draw up regionally focused strategic relationship programs that establish positive, sustained, and objective-driven military, civil, and commercial relationships in emerging and littoral nations.

Prioritize Assets for Increased Protection

Finding: Protection of all assets in all environments is not possible; naval component commanders therefore need to be able to conduct risk assessments to allow prioritization of protection measures and operational alternatives. Certain critical economic infrastructure represents maritime targets that have strategic consequence. These warrant increased monitoring as part of maritime security.

Major Recommendation 9: The Deputy Chief of Naval Operations for Information, Plans, and Strategy (N3/N5) and the Deputy Commandant of the Marine Corps for Plans, Policies, and Operations (DCMC[PPO]) should develop a comprehensive and integrated critical infrastructure protection (CIP) assessment and planning process, in cooperation with the naval component commanders, the Commandant of the Coast Guard, and other relevant offices in the Department of Homeland Security, to increase the focus on maritime areas of strategic interest or importance.

The plan would include training in the identification and risk assessment of strategic maritime infrastructure and expanded surveillance. N3/N5 should take advantage of the assessment approaches that have been screened by the Office of the Undersecretary of Defense (Policy). It should also include the deployment of foreign area officers, civil affairs officers, maritime liaison officers, and regional specialists to ensure that information on critical assets is current and accurate.[18] It will also be necessary for N3/N5 to develop a deliberate strategy to strengthen and expand the maritime liaison network, leverage DOD's commercial partners, and coordinate with the host country maritime community, embassies, the Coast Guard, and other agencies.

Operationalize the "National Fleet" Concept

Finding: Operationalizing the National Fleet means going beyond purchasing common equipment and cursory integration of training. It means connecting the Navy and the Coast Guard in real time for operations and programs. It means synergizing intelligence, command centers, forward engagement, boardings, small boat operations, tactics, and operations to take advantage of the core competencies and unique characteristics of the two Services. It means using the Coast Guard as an instrument to gain access where the Navy might not be able to engage initially. It means leveraging each Service's unique relationships with the rest of the maritime community—the Navy with other nations' naval forces, and the Coast Guard with the commercial, international, and other Coast Guard agencies. It means creating not a bright line between the two, but rather an agreed-upon overlap that will enable both to respond to the full spectrum of events, wherever they might occur.

Major Recommendation 10: The Chief of Naval Operations and the Commandant of the Coast Guard should reinforce their commitment to the National Fleet. They should direct the Navy and the Coast Guard to train together for GWOT operations to the maximum extent prudent, especially in tactics involving the new expeditionary command and boarding schools, and patrol boat operations. Current discussions proposing a global Joint Force Maritime Component Commander should be linked in terms of a National Fleet. Each Service should provide and be an active conduit for maritime information and intelligence. An expansion and funding of the National Maritime Intelligence Center is critical, as is mutual staffing and interoperability of the two Maritime Intelligence Fusion Centers.

Finding: As part of the National Fleet strategy, the Navy needs to analyze both the direction of the littoral combat ship (LCS) program and the requirement for additional smaller boat assets. The current relegation of the GWOT to the status of a mission module that includes only special warfare sea-air land teams

[18]These assets include, e.g., oil and gas exploration and production facilities, pipelines, cargo and bulk commodity terminals and wharves, sea approaches, anchorages, buoy and navigation systems, and telecommunications cables.

(SEALS) and equipment is misleading. In fact, most of the planned mission modules, for example, antisurface warfare, which includes helicopters, autonomous vehicles, guns, and mine warfare, should, and do, support GWOT missions—but the GWOT does not appear to be their primary focus. In addition to the LCS, the Navy appears to be supporting the concept of the high-speed vessel and has evaluated with the Army the use of ships to address rapid deployment and sea-basing concepts. The high-speed vessel could also be a force multiplier for the GWOT by contributing to rapid deployment, MDA, and logistical resupply for a dispersed fleet. Continued experimentation and support for this program are needed.

Major Recommendation 11: The Chief of Naval Operations and the Commandant of the Coast Guard should develop a National Fleet strategy to address the GWOT. As part of this strategy, and recognizing fiscal realities, the Navy and the Coast Guard should determine requirements for the littoral combat ship and for high-speed, small patrol vessels.

Attract, Develop, and Retain the Right People

Finding: Personnel with the right skills and motivation are by far the most critical success factor for naval forces prosecuting the GWOT, but the committee did not see that GWOT-related personnel issues were being addressed in any coherent manner. Many existing career tracks will contribute, but for the specialized needs embodied in the Naval Expeditionary Combat Command, a tailored plan and implementation are called for, especially since many of the specialists important to the GWOT are not ones historically valued by the Navy compared to fleet operational personnel.

Major Recommendation 12: The Deputy Chief of Naval Operations for Manpower, Personnel, Training, and Education (N1) should take the following steps:

- Develop a GWOT personnel support plan that will address end-to-end recruitment/accessions, training, education, assignments, rotations, and promotion opportunities, taking into account what should be complementary staffing in other Services and agencies of both DOD and other government departments and agencies.
- Initiate and sustain a foreign area officer program by recruiting trained personnel from the Navy, Navy Reserve, and assigned Marine Corps personnel with the desired qualifications, and then expand the program to include enlisted personnel.
- Initiate and sustain a training and development program for qualifying riverine operators by sourcing training support from Marine Corps and contracted Vietnam War riverine veterans, and other sources, foreign and domestic.

- Define and recruit the civil service talent needed to build an enduring knowledge base for naval forces.

N1 should also expand senior officer training to prepare for decision making in the expected complex and ambiguous circumstances of many, if not most, of the anticipated GWOT scenarios. The range of scenarios is wide and far reaching; careful development of a representative set that stresses decision making under uncertainty and time constraints is required.

Appendixes

A

Terms of Reference

At the request of the former Chief of Naval Operations, the Naval Studies Board of the National Research Council conducted an assessment of the adequacy of and prospects for improving the role of Naval Forces in the Global War on Terror (GWOT). The specific terms of reference were as follows:

- Review the Department of the Navy's draft strategy for homeland defense and assess its capability to integrate required "find and fix" capabilities, including the ability to ensure covert access from the sea, conduct short notice strike operations using kinetic and non-kinetic effects, and conduct Maritime Security Operations, including current Maritime Intercept Operations capabilities, concepts of operations, and both technical and operational limiting factors.
- Identify technologies for improving the role of naval forces in the GWOT, including sensor technologies for ships' detection and cargo inspection in Maritime Security Operations, as well as to mitigate the cargo/terrorist search problem by cargo "tagging" and handling identification.
- Review the status of databases and their management pertaining to cargo and terrorists, and assess their capabilities to communicate, share, and integrate relevant command, control, communications, computers, intelligence, surveillance, and reconnaissance data to all involved including cooperating foreign navies.
- Evaluate the interface, interoperability, and integration of current Navy and Marine Corps capabilities with the Coast Guard, other Services, and civilian, federal, state, and local agencies for the GWOT, including concepts of operations, corresponding systems, and technology requirements utilized for conducting Maritime Security Operations.
- Identify other naval-unique capabilities that can enhance the role of naval forces in the GWOT, such as non-lethal weapons as applicable to Maritime

Intercept Operations and mine countermeasures to assure access to and egress from key ports.

- Examine other factors related to the role of naval forces in the GWOT, including training technology applicable to boarding and search; responsibilities, organizational controls, and capabilities for risk assessments at various levels; definitions and validations of Maritime Intercept Operations readiness measures; possibilities of container modification to facilitate automated search of weapons of mass destruction content (e.g., valves permitting evacuation to test air); and costs and manpower issues (e.g., can boarding be done adequately with reduced crews?).
- Conduct an assessment of other potential options not currently under consideration that merit additional study. This effort should focus on methods that utilize small, distributed forces that do not require a big footprint.

B

Committee and Staff Biographies

Miriam E. John, Co-chair, is vice president emeritus of Sandia National Laboratories. Her areas of expertise include nuclear weapons development; chemical, biological, radiological, and nuclear defense; homeland defense and security; Department of Defense (DOD) acquisition and management; and asymmetric warfare capabilities. Dr. John serves on numerous scientific boards and advisory committees, including the DOD's Defense Science Board and Threat Reduction Advisory Committee, the External Research Advisory Board for the University of California, Davis, and the California Council on Science and Technology. She is a past member of the Air Force Scientific Advisory Board and the NRC's Board on Army Science and Technology. She is the acting chair of the External Advisory Board of the Chemical and Biomolecular Engineering Department of Tulane University and vice chair of the NRC's Naval Studies Board.

Richard L. Wade, Co-chair, is a principal at Exponent; he was formerly president of Risk Management Sciences, a private consulting firm specializing in risk mitigation and threat assessment. He is also adjunct associate professor of medicine at the University of California at San Francisco Medical Center and maintains an active international private practice in environmental health. Dr. Wade's expertise is in risk mitigation and threat assessment, and his career has included work as a regulator, professor, and consultant. He has served as the head of public health agencies in Seattle, Washington, and for the states of Minnesota and California and has taught at the University of Minnesota and the University of California. He has expertise in local, state, federal, and international environmental and health issues. In 1990, he received the American Public Health Association's lifetime achievement award. Dr. Wade is a member of the NRC's Naval Studies Board.

H. Norman Abramson is retired executive vice president of Southwest Research Institute. He is an expert in a wide variety of complex systems ranging from submarines to surface ships, ground transport vehicles to highways, and airplanes to spacecraft, and such massive entities as nuclear power plants and offshore drilling/production facilities. He has served on many NRC committees, of which three dealt with terrorism. He is a member of the National Academy of Engineering.

Noel K. Cunningham recently retired as director of operations for the Port of Los Angeles, where he managed the port police, port pilot, and emergency management divisions. He has also been chief of police for the Port, which has the only U.S. police organization dedicated to port activities. Mr. Cunningham's background is in homeland defense and risk assessment and in federal, state, and local laws that apply to cargo protection, pollution investigations, vessel traffic control, and drug interdiction.

Kevin P. Green retired from the U.S. Navy in 2004 with the rank of vice admiral after 33 years of service. He is employed by IBM Global Business Services as director, Department of Defense Strategic Accounts. During his Navy career, Admiral Green served as Deputy Chief of Naval Operations (Plans, Policy and Operations [now Information, Plans, and Strategy]); Commander, U.S. Naval Forces Southern Command; and Commander, Cruiser-Destroyer Group Three. His expertise includes joint operations, political–military affairs, naval operations, and Department of Navy planning, programming, budgeting, and execution.

Rodney Gregory is a project manager for the Supply Chain Management Strategy practice at IBM Global Business Services and is a member of its Defense Logistics services team. Mr. Gregory has more than 30 years of experience in the transportation and logistics industry as a professional mariner, consultant, and corporate officer with an ocean carrier. Since joining IBM in 1998 he has worked with DOD clients to help transform the business processes and systems that support their transportation and logistics activities. He was a member of the NRC's former Marine Board, which advised the government on maritime issues. He was previously appointed by the Secretary of Transportation to the Navigation Safety Advisory Council (NAVSAC), serving from 1994 until 2000. He is also a member of the National Defense Transportation Association (NDTA) as well as other industry groups.

Lee Hammarstrom is special assistant for space and information technology to the director of the Applied Research Laboratory at Pennsylvania State University (ARL/PSU). His areas of expertise include systems for command, control, communications, computers, intelligence, surveillance, and reconnaissance; and military and intelligence space systems. Before joining the ARL/PSU, he was the first chief scientist at the National Reconnaissance Office (NRO) and was chief scientist at the Office of the Secretary of Defense for Command, Control, Communications, and Intelligence. Mr. Hammarstrom has also held positions at the Naval Research Laboratory in remote sensing, reconnaissance, and intelligence that led to the creation of the space system engineering division. Before that he worked in industry for nine years. Mr. Hammarstrom was named NRO Pioneer in 2002 for his 40 years of contributions to national reconnaissance.

Paul W. Hoff recently retired as director of Advanced Sensor and Distributed Fusion Systems for BAE Systems, where he was responsible for organizing and managing leading-edge technology teams in the development of integrated fusion and advanced unattended ground sensors and unattended air vehicles supporting military transformation. He was chief technology officer for company-wide initiatives in homeland security emphasizing detection of weapons of mass destruction, and he directed advanced research programs integrating industry, government laboratories, and major universities. Dr. Hoff has experience with Army Advanced Sensors Collaborative Technology Alliances programs to develop integrated fusion and advanced unattended ground sensors and unattended air vehicles supporting military transformation.

James D. Hull retired from the U.S. Coast Guard with the rank of vice admiral. His background is in Coast Guard and interagency operations, maritime security, and maritime intercept operations and capabilities. He is an independent consultant specializing in homeland security and maritime issues and as a senior mentor supporting the U.S. Joint Forces Command's Joint Warfighting Center. He has been commander of the Coast Guard's Atlantic area and the U.S. Maritime Defense Zone Atlantic, where his immediate responsibilities included protecting the vital ports, waterways, and borders of the United States and supporting all military preparations required by the DOD as an integral member of the Department of Homeland Security when it was formed.

Harry W. Jenkins, Jr., retired from the U.S. Marine Corps with the rank of major general. He was the director of business development and congressional liaison at ITT Industries, where he was responsible for activities in support of tactical communications systems and airborne electronic warfare systems with the Navy, Marine Corps, Coast Guard, National Guard, and relevant committees in Congress. General Jenkins's background is in expeditionary warfare, particularly with regard to its mission use of command, control, communications, computers, and intelligence (C4I) systems. During Desert Storm, General Jenkins commanded the Fourth Marine Expeditionary Brigade, for which he directed operational planning, training, and employment of the ground units, aviation assets, and command-and-control systems for the 17,000-person amphibious force. General Jenkins's last position before retirement from the Marine Corps was as director of expeditionary warfare for the Chief of Naval Operations. In that position he initiated a detailed program for improving the C4I systems for large-deck amphibious ships, managed all programs of naval mine warfare, and reorganized the Navy's unmanned aerial vehicle operations from aircraft carriers and amphibious ships. He is a member of numerous professional societies, including the Marine Corps Association, Marine Corps Aviation Association, Expeditionary Warfare Division of the Naval Defense Industry Association, Navy

League, and Adjutant Generals Association of the United States. General Jenkins is also a member of the NRC's Naval Studies Board.

Ronald R. Luman is head of the National Security Analysis Department at the Johns Hopkins University's Applied Physics Laboratory. Dr. Luman has a broad base of experience in applying systems engineering principles to guidance system accuracy, unmanned undersea vehicles, countermine warfare, ballistic missile defense, and intelligence systems. He has been chief analyst for the Joint Countermine Advanced Concept Technology Demonstration, a principal in a watershed study to define the sea-based components of the ballistic missile defense architecture, and technical director for intelligence systems engineering and architecture. Most recently, he initiated an annual symposium on national security challenges posed by unrestricted warfare.

Ann K. Miller is the Cynthia Tang Missouri Distinguished Professor of Computer Engineering at the University of Missouri, Rolla. Her areas of expertise include information assurance, with an emphasis on computer and network security, and computer engineering, with an emphasis on large-scale systems engineering, satellite communications, and real-time software. She was formerly deputy assistant secretary of the Navy for research, development, and acquisition (command, control, communications, computers, and intelligence; electronic warfare; and space); Department of the Navy chief information officer; and director for information technologies for DOD research and engineering.

John H. Moxley III recently retired as managing director at North American Health Care Division, Korn/Ferry International. His background is in internal medicine; military medical issues; chemical, biological, radiological, and nuclear detection; training, costs, and manpower issues; and federal agency administration. He served as Assistant Secretary of Defense for Health Affairs during the Carter administration. Dr. Moxley is a member of the NRC's Board on Army Science and Technology and was chair for the NRC report, *Protecting Those Who Serve: Strategies to Protect the Health of Deployed U.S. Forces*. He is also a member of the Institute of Medicine and the NRC's Naval Studies Board.

Gene H. Porter is an independent consultant. His areas of expertise include matters relating to national security planning and weapon systems development and to defining defense planning scenarios to guide the development of the U.S. military force structure. Mr. Porter previously served as director of acquisition policy and program integration for the Office of the Under Secretary of Defense for Acquisition. He has also served on numerous scientific and advisory committees and chaired the NRC's Committee for Mine Warfare Assessment. He is a member of the NRC's Naval Studies Board.

John S. Quilty retired as senior vice president and director of the command, control, communications, and intelligence (C3I) DOD Federally Funded Research and Development Center at the MITRE Corporation. His background is in supporting the technical requirements of the Army, Navy, Defense Information Systems Agency, Office of the Secretary of Defense, Office of the Joint Chiefs of

Staff, and other members of the national security community. Mr. Quilty's recent work focused on support of DOD initiatives and activities to improve command, control, communications, computer, and intelligence support to joint operations. He is a member of the NRC's Naval Studies Board.

Brad Roberts is a research staff member in the Strategy, Forces, and Resources Division at the Institute for Defense Analyses. His areas of expertise include both macroscale systems and policy and weapons of mass destruction and counterterrorism. He is a member of the DOD's Threat Reduction Advisory Committee and has been a research fellow at the Center for Strategic and International Studies. He is also an adjunct professor at the George Washington University Elliott School of International Studies; chair of the research advisory council of the Chemical and Biological Arms Control Institute; and a member of the International Institute for Strategic Studies and the Council on Foreign Relations.

Annette L. Sobel is a distinguished member of the technical staff at Sandia National Laboratories. Her areas of expertise include advanced technology development and unconventional threat analysis in support of chemical and biological countermeasures and in the field of human factors (e.g., critical decision making under stress). She is a brigadier general in the U.S. Air Force and the director of intelligence for the National Guard Bureau. Dr. Sobel's military experience includes combat and chemical-biological warfare medical response unit commands. She serves on the Defense Intelligence Agency's Advisory Board.

H. Eugene Stanley is a professor of physics and director of the Center for Polymer Studies at Boston University. His areas of expertise are polymers and sensors, theory of phase transitions and critical phenomena for a wide range of systems including the XY and n-vector models, percolation theory, water structure, and application of statistical mechanics to biology, economics, and medicine. Dr. Stanley's current research interests deal with understanding threat networks and threatened networks. He is a member of the National Academy of Sciences.

Marlin U. Thomas is dean at the Graduate School of Engineering and Management, Air Force Institute of Technology. He previously was professor of industrial engineering at the Purdue University School of Industrial Engineering. His area of special expertise is logistics systems for contingency operations. He has been president of the Institute of Industrial Engineers (IIE) and was a member of the Army Science Board. Currently, he is associate editor for *IIE Transactions* and *Computers and Industrial Engineering*. Dr. Thomas is a fellow of the American Society for Quality, Institute of Industrial Engineers, and Institute for Operations Research and Management Sciences. He retired from the U.S. Navy Reserve, Civil Engineer Corps, with the rank of captain.

David A. Whelan is vice president-general manager and deputy president of the Boeing's Phantom Works Division. His areas of expertise include defense research and development, such as space systems, tactical military systems, unmanned vehicles, and space-based moving target indicator radar systems.

Before joining Boeing, he was director of the Tactical Technology Office at the Defense Advanced Research Projects Agency. Dr. Whelan is a member of the NRC's Naval Studies Board.

Elihu Zimet is a distinguished research professor at the Center for Technology and National Security Policy at the National Defense University (NDU). His background is in naval science and technology related to military operations, kinetic and nonkinetic effects, and low-observable and counter-low-observable technologies. Before joining NDU he was head of Special Programs and of the Expeditionary Warfare Science and Technology Department of the Office of Naval Research. Dr. Zimet is a member of the NRC's Naval Studies Board.

Staff

Charles F. Draper is director of the NRC's Naval Studies Board. Before joining the NRC in 1997, Dr. Draper was the lead mechanical engineer at S.T. Research Corporation, where he provided technical and program management support for satellite Earth station and small-satellite design. He received his Ph.D. in mechanical engineering from Vanderbilt University in 1995; his doctoral research was conducted at the Naval Research Laboratory (NRL), where he used an atomic-force microscope to measure the nanomechanical properties of thin-film materials. In parallel with his graduate studies, Dr. Draper was a mechanical engineer with Geo-Centers, Incorporated, working at NRL on the underwater X-ray backscattering tomography system to be used for the nondestructive evaluation of U.S. Navy sonar domes on surface ships.

Arul Mozhi is senior program officer at the NRC's Naval Studies Board; he has also served as senior program officer at the NRC's Board on Manufacturing and Engineering Design and National Materials Advisory Board. Before joining the NRC in 1999 Dr. Mozhi was senior scientist and program manager at UTRON, Inc., working on applying pulsed electrical and chemical energy technologies to materials processing. From 1989 to 1996 Dr. Mozhi was a senior engineer and task leader at Roy F. Weston, Inc., a leading environmental consulting company, working on long-term nuclear materials behavior and systems engineering related to nuclear waste transport, storage, and disposal in support of the U.S. Department of Energy. He had previously been a materials scientist at Marko Materials, Inc., working on rapidly solidified materials. He received his B.S. in metallurgical engineering from the Indian Institute of Technology and his M.S. and Ph.D. degrees in materials engineering from the Ohio State University, where he was also a postdoctoral research associate.

C

Summary of Committee Meeting Agendas

The Committee on the Role of Naval Forces in the Global War on Terror first convened in July 2005 and held additional meetings and site visits over a period of 7 months:

- *July 21-22, 2005, in Washington, D.C.* Organizational meeting: Office of Chief of Naval operations briefings on Navy investment and capability additions to the force, operations perspective, requirements overview, fleet readiness, and logistics perspective; Office of the Assistant Secretary of Defense for Homeland Defense briefing on maritime homeland defense in the global war on terror; and U.S. Coast Guard and Headquarters and U.S. Marine Corps briefings on their perspective on the role of naval forces in the global war on terror.
- *August 18-19, 2005, in Washington, D.C.* Office of Chief of Naval operations briefings on studies, war games, and campaign analysis related to the global war on terror and on the role of naval riverine forces; U.S. Northern Command briefing on homeland defense and maritime domain awareness; Office of Naval Intelligence briefing on threat assessment of the maritime domain; CNO Strategic Studies Group briefing on transforming maritime forces for the global war on terror; and CNA Corporation briefing on naval assessments on the global war on terror.
- *September 26-27, 2005, in San Diego, California.* Naval Special Warfare Command briefings on organizational and operational relationships with SOCOM and the Navy, major programs, training, and associated equipment, present and future challenges, current operations and relevant past global war on terror operations, and explosive ordnance capabilities and marine mammal program; U.S. Pacific Fleet and U.S. Marine Corps Forces, Pacific, briefings on their roles in

the global war on terror; Space and Naval Warfare Systems Center and Naval Postgraduate School briefings on their initiatives in the global war on terror.

- *November 1-2, 2005, in Norfolk, Virginia.* Fleet Forces Command, Joint Forces Command, U.S. Second Fleet, U.S. Third Fleet, Naval Network Warfare Command, Navy Expeditionary Combat Command, Atlantic Submarine Force, Atlantic Naval Air Force, Atlantic Surface Force, and Coast Guard Atlantic Area briefings on their roles in the global war on terror.
- *November 30–December 2, 2006, in Suitland, Maryland, and Washington, D.C.* Maritime Domain Awareness Program Integration Office (U.S. Coast Guard), National Maritime Intelligence Center (Office of Naval Intelligence), National Reconnaissance Office, and Naval Research Laboratory briefings on their capabilities and initiatives in the global war on terror; U.S. Special Operations Command briefing on its role in the global war on terror; Office of Chief of Naval operations briefing on an update of its global war on terror initiatives; National Defense University briefing on alternative fleet architectures study; World Shipping Council briefing on understanding the shipping container security problem; and John Hopkins University School of Advanced International Studies, Congressional Research Service, and Center for Strategic and Budgetary Assessments panel discussion on adapting to unconventional/unrestricted warfare.
- *December 20-21, 2005, in Washington, D.C.* Defense Advanced Research Projects Agency and Office of Naval Research briefings on their programs and initiatives in the global war on terror.
- *January 23-27, 2006, in Irvine, California.* Committee deliberations and report drafting.

D

Acronyms and Abbreviations

ASN(RDA)	Assistant Secretary of the Navy for Research, Development, and Acquisition
C4ISR	command, control, communications, computers, intelligence, surveillance, and reconnaissance
CIP	critical infrastructure protection
CNO	Chief of Naval Operations
COCOM	combatant commander
CONOPS	concept of operations
CONUS	continental United States
DHS	Department of Homeland Security
DOD	Department of Defense
EOD	explosive ordnance disposal
GWOT	Global War on Terror
ISR	intelligence, surveillance, and reconnaissance
LCS	littoral combat ship
MCCDC	Marine Corps Combat Development Command
MCO	major combat operation
MDA	maritime domain awareness
N1	Deputy Chief of Naval Operations for Manpower, Personnel, Training, and Strategy
N3/N5	Deputy Chief of Naval Operations for Information, Plans, and Strategy
N8	Deputy Chief of Naval Operations for Integration of Capabilities and Resources
NOAA	National Oceanographic and Atmospheric Administration

NORTHCOM	U.S. Northern Command
NRC	National Research Council
NWDC	Navy Warfare Development Commant
OPNAV	Office of the Chief of Naval Operations
QDR	Quadrennial Defense Review
SEAL	sea-air-land (team)
USMC	U.S. Marine Corps
USN	U.S. Navy